Secrets of Mental Math

Master the Art of

Advanced Calculation and Memorization All in your Head

Kenneth Paek

All rights reserved. No part of this publication may be reproduced in any form without written consent of the author and the publisher. The information contained in this book may not be stored in a retrieval system, or transmitted in any form by any means, electronic, mechanical, photocopying or otherwise without the written consent of the publisher. This book may not be resold, hired out or otherwise disposed by way of trade in any form of binding or cover other than that in which it is published, without the written consent of the publisher. Respective authors own all copyrights not held by the publisher. The presentation of the information is without contract or any type of guarantee assurance. All trademarks and brands within this book are for clarifying purposes only and are the owned by the owners themselves, not affiliated with this document.

Disclaimer

The information contained in this book is strictly for educational purpose only. The content of this book is the sole expression and opinion of its author and not necessarily that of the publisher. It is not intended to cure, treat, and diagnose any kind of disease or medical condition. It is sold with the understanding that the publisher is not rendering any type of medical, psychological, legal, or any other kind of professional advice. You should seek the services of a competent professional before applying concepts in this book. Neither the publisher nor the individual author(s) shall be liable for any physical, psychological, emotional, financial, or commercial damages, directly or indirectly by the use of this material, which is provided "as is", and without warranties. Therefore, if you wish to apply ideas contained in this book, you are taking full responsibility for your actions.

Table of Contents

1: Learning the Tricks of Quick and Impressive Calculations

2: The Mental Game of Subtraction and Addition

3: Intermediate Multiplication: Moving beyond the Basics

4: Mastering the Magic of Mental Division

5: Learning How to Make Ballpark Estimates

6: Speeding up Pencil and Paper Calculations

7: Perfecting the Art of Memorizing Numbers

8: Stretching Your Mind with Advanced Multiplication

9: Understanding the Magical Secrets of Arithmetic

Conclusion

Important Insight

Mathematics is not a new subject to us because almost everyone knows how to count in whichever language we are comfortable with. Math is not only the game of numbers but also the language of nature.

As a matter of fact, it has been established through research that more we understand our universe; the more we discover interesting mathematical connections. For instance, flowers have spirals that are carefully line up in a special sequence of numbers commonly referred to as Fibonacci numbers. This sequence can be studied, understood and generated. A study on seashells also reveals that they form perfect mathematical curves known as logarithmic spirals that originate from a chemical balance. Apart from the flowers and seashells, star clusters also form a mathematical arrangement in the way they tag on one another. This has been observed through astronomical studies.

Scientists including mathematicians have spent centuries trying to discover the nature of math and how to develop a mastery of the same. It is possible to get comfortable with calculations of any nature as long as you discover the secrets behind the numerics.

The reason why it is important to unravel and enhance your command of math is because numbers are part of our daily lives. For example, if you need to call someone, you need a phone number. The time you spend on your phone conversation is measured in hours and minutes which are also numbers. Think of any date in history including your birthday, graduation day or just any other significant day and you will find that recording it in numbers is much easier than in any other form.

This book will show you the tricks and shortcuts that will make math so intriguing. You will definitely impress your friends and colleagues with your lightning fast calculating ability. But most importantly, these secrets will benefit you at a personal level because you will be able to handle complex numeric problems within seconds.

1: Learning the Tricks of Quick and Impressive Calculations

When dealing with numbers, speed is of essence. You need to have a quick grasp of the problem and manipulate the digits so it will give an accurate answer almost immediately. This is what will make you standout. Oftentimes, mathematics is portrayed as a subject with rigid rules that leave little if any space for creative thinking. As you will learn in the following chapters, there are several techniques you can use to solve the same problem.

Solving larger problems can be done quickly by first breaking down the numbers into smaller and more manageable components. The devil is in the details! When you understand the special features of math problems, you will find it easier to solve them.

The Secrets behind Instant Multiplication

Learning how to multiply digits in your head is probably the first breakthrough that will lift your level of curiosity and self-esteem. To show you an example of how easy multiplication can be once you know the secret, consider the below math problem:

52 x 11

To solve this problem and get an accurate answer, add the digits 5 and 2 and put the answer between the 5 and the 2. There you have it! 572

You can also try 63 x 11

Add 6 and 3 which will give you 9. Then put the 9 between 6 and 3 and this will give you the answer 693.

To jog your mind without writing on a piece of paper, try 71 x11. If your answer is 781, then you are on the right track to unraveling the secret of instant multiplications.

In the above calculations, you may have noticed that the two numbers you have been adding do not exceed 9. However, there are situations where the sum of the two numbers exceeds 9. In this case, you will have to use a slightly different approach as shown below:

Assume you have the following problem:

76 x 11

When you add 7 and 6, you will get 13. Now, with the above approach you will have 7136 which is **NOT** correct. From calculations, you know the answer is supposed to be 836 but the question is - how do you arrive at that? It is quite simple:

In the summation, you have two digits 1 and 3. Take the 3 and put it between 7and 6 and take 1 and add it to 7 so you get the correct answer of 836.

To cement your understanding, let us try out 68 x 11. Adding 6 and 8 gives you 14. Put 4 between 6 and 8 and add the 1 to 6. The answer should be 748.

To test to see if you have the right concept, try 97 x11.

If you answer as 1067, then you are inching closer to becoming a mathemagician.

After the first round of multiplication magic, you may have some questions regarding the multiplication of 3 digit numbers, or even larger numbers then 11. This is a genuine concern since the above problems were all two digits in nature.

To solve this query, let's look at the problem 325 x 11. Your answer should still begin with 3 and end with 5.

Here is how you solve it.

Take 3 + 2 = 5 and 2 + 5 = 7. Then, go ahead and put the 5 and the 7 between 3 and 5. In essence, you are replacing the center digit with 57 giving you the answer 3575.

If the additions exceed 9, this is how to handle it:

Assume you have the following problem to solve; 497 x 11. 4 + 9 = 13 take the 1 and add to 4 giving you 5 and leave the 3 alone. Then 9 + 7 = 16, take the 1 and add to 3 giving you 4 and leave the 6 as it is.

Therefore the center number that will replace the 9 in 497 is 46. The first digit, 4 will change to 5 because of the addition above. The answer will therefore be 5467.

Squares of Numbers

Apart from the multiplication by 11 above, consider the case of squares. This is where a number is multiplied by itself such as 8 x 8, 6 x 6 and so on. Let us first look at numbers that end in 5.

In order to find the square of a two digit number that ends in 5, remember the following two important things:

The answer starts with the multiplication between the first digit and the next higher digit.

Your answer should end in 25

Take for instance, the square of 35. Start by multiplying the first digit which is 3 by the next higher digit that is 4. Thereafter, attach 25 at the end. This should give you 3 x 4 = 12, and then attach 25 and your answer becomes 1225.

For the case of 65, take 6 x 7 = 42, then attach 25 to give you 4225.

NB: You can employ the same trick when you are multiplying two digit numbers that have similar first digits and with second digits that add to 10. In computing your answer, take the first digit and multiply by the next higher digit, and then attach to this answer the product of the second digits. Take for instance, 97 x 93. 9 is similar and 7 + 3 = 10.

This will give you 9 x 10 = 90 then attach 7 x 3 = 21 to get 9021.

You can also try 46 x 44

4 x 5 = 20 and then 6 x 4 = 24; your answer should be 2024.

In the coming chapters, you will learn how to work out multiplications where the last digits do not add to 10.

2: The Mental Game of Subtraction and Addition

The tradition, when it comes to adding and subtracting numbers is the right to left rule. This method, despite its popularity in schools makes addition and subtraction a slow process that may at times give erroneous solutions. This chapter introduces the direct opposite of left to right rule of solving math additions and subtractions. This will help you call out answers to problems well ahead of the rest. As a matter of fact, you won't even need a pencil.

Left-Right Addition

If you are doing math on paper, the right to left method of addition and subtraction may work well for you. However, if you want to break away from the ordinary and do the math all in your head at lightning speed, then you should change the methodology and work from left to right.

The reason why this is simple is because all numbers are pronounced from left to right such as fifty four (54), forty eight (48) and eighty seven (87). Following the same grain of pronunciation, you should start calculating numbers from left to right.

Whenever you compute any problem from right to left, the answer is generated backwards. This is why it becomes so difficult to do math in your head in the right to left fashion.

Working from left to right on the other hand will help you start with the most significant digits of your calculation. With practice, you can completely master the left to right methodology and this will go a long way into making your mental calculations natural and efficient.

Two Digit Additions

The assumption here is that you already understand and have no problem with single digit additions and subtractions. Two digit additions are also fairly easy to do in your brain, but most importantly they will introduce you to a concept to handle larger problems. The two digit addition concept also introduces one fundamental principle in mental arithmetic where you simplify your problem by breaking it up into smaller and manageable parts.

The easiest two digit calculations involve problems that do not need to carry on any numbers. This means that the first digits sum up to 9 while the last digits also add up to 9 or below. For instance, the problem 37 + 22 can be solved by first adding 37 to 20 then add 2. After the addition of 20, the problem becomes simpler because all you need is to add the resulting 57 to 2 to give you 59.

37 + 22 = 37 + 20 + 2 = 59. This is the same process the mind goes through in arriving at the answer.

When the calculation requires that you carry a number, here is how to approach it:

47 + 38. Add from left to right 47 + 30 = 77. Then add the 77 to 8 making it 85.

47 + 38 = 47 + 30 + 8 = 85.

On your own, you can try calculating the following problems mentally:

    a. 94 + 57
    b. 64 + 43
    c. 89 + 78

The above explanations represent a systematic manner of handling mental calculations. It may take you a bit of time to grasp the concept, but once you have internalized it, you will begin to hear and see these numerics right in your mind.

Additions Involving Three Digits

The concept used here is similar to the technique used in adding two digit numbers. You will also have to add from left to right because that is how the brain works. You will discover that after each step in your addition process, you arrive at a new and much simpler problem. Let's try out the following problems:

437 + 698 (600 + 90 + 8)

The diagrammatic illustration of the mental process is as follows:

437 + 600 = 1037 then 1037 + 90 = 1127 and lastly 1127 + 8 = 1135

Any kind of mental addition can simply be done with this method. The secret here is to keep on simplifying the problem until you end up with one digit number. You may have realized that adding

437 to 698 requires that you hold 6 digits in your head. Whereas adding 1127 to 8 only require you to hold 5 digits. This makes the problem easier.

In 632 + 174 (100 + 70 + 4), you will find you first add the hundreds 632 + 100 = 732. Then add the tens in 732 + 70 = 802. Then finally add the ones giving you 802 + 4 = 806.

When doing these problems, you should first practice out loud. The verbal reinforcement will help you learn the mental method faster.

When adding three digits which are nearing round figures, for instance, 497, 698, and 599 and so on, the best way is to round up to the nearest hundred and then subtract the difference between the three digit number and the rounded figure after doing the summation.

Check this example:

749 + 597 (500 + 90 + 7); you can approach this problem in two ways. First, you can add 749 to 500 to give you 1249 then add 1249 to 90 which will give you 1339 and lastly add 1339 to 7 to give you 1346.

This seems a long process as compared to 749 + 600 then subtracting 3 from the sum.

The reason why you break the second number is purely conventional and does not have mathematical significance. You can also break up the first number if that works for you.

When adding three-digit to four-digit numbers, most people encounter coordination problems. This is understandable because most human memories can hold up to 7 or 8 digits at a time without resulting to calculators and other artificial memory devices. In many problems, the last digit in one or both of the numbers being added end in a zero such as 2700, 310, 420.

Assume you have the following problem:

1400 + 523; since 14 hundred + 5 hundred is 19 hundred, you simply attach the 23 to 19 hundred to get 1923.

Left to Right Subtraction

When compared to addition, subtraction may be little more difficult. However, with left to right computation, and breaking down the problem into simpler parts, subtraction becomes as easy as addition.

Two Digit Subtractions

Here, the goal is to simplify and to reduce the computation by subtracting a single digit number. The following is an example of this concept:

76 – 25 (20+5). First step, you subtract 20 from 76, thereby ending up with 56, then further subtracting 5 to get your answer, which is 51.

76 – 25 = 56 – 5 = 51

When you encounter numbers nearing round figures such as 38, 29 and 69, the best way is to handle them as follows:

In 75 – 38, you can subtract 40 from 75, and then add 2. This is much easier than subtracting 30 from 75, then subtracting 8 from the answer.

The rule for whether you decide to do a straight subtraction or a mixture of subtraction and addition is this; if a two digit subtraction requires borrowing, the best approach is to round up the second number to a multiple of 10, then subtract the rounded number, then add back the differential.

This same line of thinking can be used when working with 3 digit subtractions.

Using Complements in Subtraction

In order to find out how far a two digit number is far from 100, the secret is the first digits should add to 9 and the last digits should add to 10. For instance, the complement of 52 must start with 4 to make the sum of the first digits 9 and then 8, which combine with 2 to form 10. So the complement is 48. The complements are determined form left to right.

Complements are very instrumental in mental subtraction because they allow you to convert complex subtraction problems into simple and straightforward addition problems.

Take for intake, 725 – 478 (500 - 22)

The first step is to subtract 500 from 725 instead of 468. This will bring you to 225. The next step is to figure out how much you should add back because you have subtracted more than the stipulated figure. This is where you need a complement. The complement of 78 above is 22. Therefore, you are adding 22 to 225 to give you 247 as your final answer.

Try to practice with many problems as you can so you may develop familiarity with the concepts described in this chapter.

3: Intermediate Multiplication: Moving Beyond the Basics

If you have ever had the opportunity to perform mathematical computations before a live audience, you know exactly how exciting it can be. Famous mathematicians such as Zerah Colburn (1804-1839) began at an earlier age of six to do lightning calculations before he could even read or write. You too may be able to impress your teachers and friends by performing mental multiplications that will perplex their minds.

Multiplication of Two Digit Numbers

There are lots of different methods you can use when multiplying two digit numbers and all of them will bring you down to the same answer.

The Addition Method

This method is used in multiplying any two digit numbers using the 2-by-1 approach. Thereafter, the results are added together. Have a look at this:

56 x 32 (30 + 2); this can be solved by taking 30 x 56 = 1680 then 2 x 56 = 112.

If you add the two together, you get 1680 + 112 = 1792.

The secret is break up 32 into 30 and 2 which are numbers you can easily multiply. Thereafter, multiply the 30 by 56 which is simply 3 x 56 with a zero attached at the end. To this product, you add the result of 2 x 56 and there you will have your answer.

Another way to handle the same problem is as follows:

56 (50 + 6) x 32; the solution will be 50 x 32 = 1600 then 6 x 32 = 192. To get your final answer, add the 1600 to 192 to get 1792.

In deciding the number to break up, the rule of thumb is to pick the number with the smallest last digit. It is easier to break up and multiply 32 than 56.

NB: 1. Numbers that end in one are always easy and attractive to break up.

    2. Where both numbers end in the same digit such as 34 and 64, you should break up the bigger number.

3. Where one of the numbers is much larger than the other as in 98 and 24, it makes a lot of mathematical sense to break up the larger number.

4. When multiplying a number that is in the fifties such as 56 with an even number like 42, it is easier to break up the number in the fifties.

The Subtraction Method

This method is particularly useful when you are dealing with numbers that end with 8 or 9. Check out the following illustration:

69 (70 - 1) x 21; this will give you 21 x 70 = 1470 and -1 x 21 = -21. Your final answer will be 1470 – 21 = 1449.

Most people find subtractions a little bit harder compared to additions. The secret here is that small numbers are easier to subtract compared to adding big numbers.

Let us look at another problem so we can clarify and simplify the subtraction method.

79 (80 - 1) x 54; 80 x 54 = 4320 and 54 x -1 = -54; your answer will be 4320 – 54 = 4266.

The emphasis here is that working out these problems right in your head will have a much significant impact than looking at the illustrations. Go through your problems one by one and pronounce the steps loudly in order to reinforce them in your thoughts.

The subtraction method is not only appropriate for numbers that end in 8 or 9 but it is also applicable for numbers that are in their high 90s such as 96, 97, 98 and 99.

Whenever the subtraction component of a math problem requires that you borrow a number, you can use the complements discussed in the previous chapter to help you solve it faster. For instance, when subtracting 78 from 330, the answer is automatically in the 200s. The difference between 30 and 78 is 48. Now take the complement of 48 which is 52. Your answer is therefore 252.

This method can be used with any type of subtraction that requires you to borrow a number. Complements are one of the most powerful tools in mathemagics. If you master this technique, you will amaze many people.

The Factoring Method

If you do not want to subtract or add any numbers while doing your multiplications, then the most appropriate method is the factoring method. It is applicable where one of the numbers in the multiplication can be factored into one digit numbers.

Factoring is simply breaking down a number into one digit numbers when multiplied together will lead you to the original answer. Take for instance, the number 24. It can be factored into 2 x 12, 8 x 3 or 6 x 4. Other examples of factored numbers include:

58 = 2 x 29

66 = 2 x 33, 3 x 22 or 6 x 11

84 = 2 x 6 x 7 or 7x 4 x 3

To see an example of how factoring can help in multiplication, look at the following problem:

56 x 42 (6 x 7); using the earlier concepts, the approach should be (56 x 40) + (56 x 2) = 2240 + 112 = 2352.

Using the factoring method, you treat 42 as 6 x 7. Begin by multiplying 56 x 6 = 336 then 336 x 7 = 2352.

The advantage of the factoring method is that it simplifies two by two multiplication problems such as 56 x 42 into easier-to-solve 3 by 1 problems such as 336 x 7 or at times 2 by 1 problems. This way, you do not have to hold a lot of numbers in your memory. Take another example 65 x 72. This can be solved as 65 x (8 x 9) = (65 x 8) x 9 = 520 x 9 = 4680. The absence of additions and subtractions makes the factoring method easier in your memory.

Whenever you come across challenging multiplication problems, try as much as possible to identify friendly products so that it can help you solve the problem faster. Friendly products are those with a zero at the center such as 301, 402, 104, 207 etc. This is how you approach friendly products.

Assume the problem 89 x 72; Break it down through factoring as follows:

89 x 72 = 89 x 9 x 8 = 801 x 8 = 6408. The friendly product 801 is easier to multiply by 8 than 712 x 9.

Three Digit Squares

The mastery of three digit squares is one of the most impressive mental feats. Just as you did with the two digit squares, you also need to round up or down to the nearest multiple of 100.

For instance in 194, you need to add 6 to get to 200. Subtracting the same 6 from 194 gives you 188. This means the problem has been reduced to 3 by 3 multiplications of 200 and 188.

You can now get the square of 194 easily by; (200 x 188) then add the product to the square of the difference (6).

Further illustrations include; square of 706

Round down 706 to 700 the difference is 6. Then add the 6 to 706 giving you 712.

(712 x 700) + 62 = 498, 436

When you encounter a problem such as 3592 you can easily do it in two steps. Add 41 to get 400 and subtract 41 to get 318, then multiply 400 x 318 to get to 127, 200. Instead of adding 127, 200 directly with the square of the difference (41), you can break up 412 as follows:

41 − 1 = 40 and 41 + 1 = 42. Multiply 42 by 40 to get to 1680, then add the square of 1 which is simply 1. The overall solution will be 127, 200 + 1680 + 1 = 128,881. Hooray!

There you have it, the square of 359. Practice with many figures as possible to master the concept.

4: Mastering the Magic of Mental Division

Mental division is an indispensable skill in our day-to-day social and business lives. Mastering this skill will save you a lot of inconveniences because every time you need to do a calculation, you will no longer need a calculator. The left to right method of computing numbers is almost natural in division.

One Digit Divisions

When doing these calculations, the first step is to figure out the number of digits that will be in your solution. For instance, in 316 ÷ 7, our interest is in the number X such that 7 x X will give us 316. From here, we know that 316 lie between 7 x 10 = 70 and 7 x 100 = 700.

The unknown value of X must therefore be between 10 and 100 meaning that our answer is a two digit number. The next step is to determine the largest multiple of 10 when multiplied by 7 gives a product that is below 316; the multiple is 40 because 50 gives us 350 which is way above 316. The answer therefore lies in the forties.

The next step is to subtract 280 from 316 giving you 36. The problem has now been simplified into a division problem that is 36 ÷ 7. From simple calculations, we know that 7 x 5 = 35 which is 1 away from 36. This gives us our final answer which is 40 + 5 + 1/7 = 45 1/7.

The above answer is a two digit and is much easier to solve. Now, consider a division problem such as 949 ÷ 4 which gives a 3 digit answer.

First, we know that 4 x 100 = 400 and 4 x 1000 = 4000. The answer therefore lies between 100 and 1000 and it is a three digit number. Thereafter, look at the multiples of 100 that bring you closer to 949. This should be 200 because 200 x 4 = 800 and 300 x 4 = 1200. Subtract 800 from 949 leaving you with 149.

Now the new division we just simplified is 149 ÷ 4. Since 4 x 30 = 120, we can base our next number at 30. Thereafter, take 149 – 120 = 29. Dividing 29 by 4 will give you 7 with a reminder of 1. The overall answer is therefore 200 + 30 + 7 + ¼ = 237 ¼.

Two Digit Divisions

The tricky part with divisions is they get harder as the divisor gets larger. To go round this one and make it simpler, consider the following problem.

598 ÷ 13; the answer (quotient) lies between 13 x 10 and 13 x 100. The first step is to ask yourself the number of times 13 goes into 590. Since 13 x 40 = 520, your answer lies in the forties. Subtract 520 from 598 to get 78. This further reduces your problem to lesser digits that is 78 ÷ 13. Taking 13 x 6 = 78, your answer is 40 + 6 = 46.

Common Tricks in Division Problems

To help you ease the strain and relax your brain when doing mental divisions, consider the following tips:

1. If the two numbers in the problem are both even, you can simplify the division by dividing each number by two before starting. For instance, in 426 ÷ 6, use the common factor two to reduce the problem to 213 ÷ 3. If the problem can be further divided by two, it will be much easier to use the factor 2 than dividing the original numbers by 4.

2. Where both numbers end in zero, divide each by 10.

3. If the numbers end in 5, double them and divide by 10 so as to make the problem simpler. For example in 35 divide by 25, double both numbers to 70 and 50 and then divide them by 10 to get 7/5 = 1 2/5.

4. If the divisor ends in 5 while the number you are dividing into has a zero at the end then multiply both numbers by two and then divide by 10.

Working out Decimalization Just like a Calculator

Converting fraction to decimals can be a breeze if you know the little secrets enshrined in the fractions from halves all the way to elevenths. Below is an easy way to memorialize:

The halves are equivalent to 0.5, the thirds to 0.333, the quarters to 0.25, the fifths to 0.20, the sixths to 0.167, the eighths to 0.125, the ninths to 0.111; the tenths are much simpler at 0.1 and the elevenths at 0. 0909.

The sevenths are a remarkable fraction. The moment you memorize 1/7, the rest of the sevenths become easy to compute. A seventh is equivalent to 0.142857 recurring. The rest of the sevenths are as follows:

2/7 = 0.285714

3/7 = 0.428571

4/7 = 0.571428

5/7 = 0.714285

6/7 = 0.857142

If you have been keen enough, you must have realized a pattern of numbers that keep on repeating themselves in each decimal derivative. Only the starting points vary.

It is easy to figure out the starting point by doing a simple and quick multiplication of the numerator by 0.14. Take for instance 2/7, the decimal equivalent starts with 0.14 x 2 = 0.28. Pick the sequence that begins with two which is 0.285714. For 3/7, it starts with 3 x 0.14 = 0.42. Pick the sequence starting with four which is 0.42857. When you come to 4/7 you will get 4 x 0.14 = 0.56.

Pick the sequence starting with 5 which is 0.571428.

The secret is to pick the sequence that starts with the digit immediately after the decimal point.

Divisibility Tests

In order to solve division problems quickly, knowing how to tell whether a number is divisible by another is critical. The following are some of the rules of the divisibility tests.

- A number is divisible by two if the last digit is even.
- To check whether a given number is divisible by 4, look at the last two digits if they are divisible by 4. Any numbers having zeros as the last two digits is also divisible by 4.
- Any number divisible by 2 and 4 is divisible by 8.
- If the sum of the digits of a number is divisible by 3 then the number is divisible 3.
- A number is divisible by 6 if it is divisible by both 2 and 3.

- For numbers divisible by 9, the sum of the digits that make up the number should be a multiple of 9. All numbers divisible by 9 are also divisible by 3 but the reverse is not true.
- A number ending with 5 or 0 is automatically divisible by 5.
- To establish whether a number is divisible by 11, you should arrive at either a multiple of 11 or zero when you alternately add and subtract the digits of the number. Take for instance 78958; $7 - 8 + 9 - 5 + 8 = 11$ hence the number is divisible by 11.

With the concepts discussed in this chapter, the science of mental division is easy to learn. As always, you need a lot of practice.

5: Learning How to Make Ballpark Estimates

As much as math involves accuracy and precision, there are some circumstances where an estimate will also serve just fine. This is especially applicable when you are receiving information in bits and pieces or from a wide range of sources and all you need is just a rough figure to guide you through. This chapter discusses guesstimation methods that you can easily use to get your way around figures whether it is multiplication, division, addition or subtraction.

Addition Estimates

In order to get a close estimate when adding large numbers, the trick is to either round up or down the original numbers. For instance, when adding 8367 to 5819, you can round them up and down as follows; 8367 to 8000 and 5819 to 6000 giving you 14,000 as the estimate figure. Since the exact answer is 14186, the error is relatively small. To be more accurate, you can round up 8367 to 8400 and round down 5819 to 5800. This will give you 14200, which is just 14 off the exact answer.

When you round off to the nearest hundred, your answer will always be off by a figure that is less than 100. Try as much as possible to keep the margin of error within 1 percent of the precise answer.

Subtraction Estimates

The logic here is the same as that in addition above. You either round up or down to the nearest tens, hundreds or thousands depending on the size of the figure. Bigger figures require higher rounding and vice versa.

Taking the same example as above, 8367 – 5819, the rounding can either be done as 8000 – 6000 giving 2000 or 8400 – 5800 giving an approximate value of 2600. The actual figure is 2548 meaning rounding off to the nearest hundreds gives you a more accurate estimate.

Division Estimates

The most important step in making division estimates is to determine the size and the magnitude of the answer. Thereafter, you can proceed with the round offs so as to make the problem easier to solve.

Take an example of 57687 divide by 6. The accurate answer is 9614.5. When you round off 57687 to the nearest thousands you get 58000. This is a much simpler figure to divide by 6 to give you 92/3 thousands which can also be expressed as 9667.

The most critical step in the problem above is to know where to place the 9. If you take 6 x 90 you get 540 and 6 x 900 gives you 5400 which are smaller figures compared to 57687. But when you take 6 x 9000, you get 54000, which is much closer to the answer. To estimate this, subtract 54 from 58 to give you 4. Then bring down the zero and divide 6 into 40 giving you 6.6666 recurring. Since you know the position of 9 as thousands, you can quickly fit in the rest of the figure like this; 9000 + 667 to give you 9667.

This form of division looks a bit simpler because the numbers involved are not as large. However, when dealing with large divisions such as 5,000,000 divide 365, you need to be very keen. To determine the magnitude of the answer, start by multiplying 365 by multiples of hundreds and thousands.

The closest you can get is 365 x 10, 000 giving you 3, 650, 000. Thereafter, round down the figure to 3, 600,000. Then to simplify the problem, divide 5, 000, 000 by 3, 600,000 which will bring you to 50 divide by 36. This gives you 17/18 which is equivalent to 1.4 since 18 goes into 70 around 4 times. So your answer is 14, 000 which is closer to the exact answer at 13698.63.

Multiplication Estimates

In doing multiplications, you also need to round up or down to simplify the problem. The best way is to round up both numbers with the same amount but in the opposite direction.

For instance, if you are multiplying 88 and 54, the best way to round them is to make them 90 (88+2) and 52(54-2). This will give you a more accurate estimate of 4680 compared to the accurate answer of 4752.

When dealing with large numbers in multiplication estimates, you should always round up the large numbers and round down the smaller numbers they are multiplied with. This will give you a lower and more accurate estimate than when you do the opposite.

Estimation of Square Roots

A square root is the number if multiplied by itself gives a value X which is its square. In most engineering and science problems, squares and square roots are common problems. Learning how to give reliable estimates of square roots will save you a lot of trouble in working out science problems. The square root of most numbers contains a decimal point or fraction.

Starting with a number as simple as 19, you can find its square root by estimating a number Y which if multiplied by itself gives a product closer to 19.

The two numbers in question are 4 and 5. Since 5 gives a product higher than 19 and 4 gives a product lower than 19, then the answer must be lying between 4 and 5.

Now divide 19 by 4 which give you 4.75. The solution, therefore lies between 4 squared and 4.75 squared. A good guess can be the center point of 4 and 4.75 which is 4.375.

Another way to look at this is as follows:

Since 4 x 4 = 16 and 19 -16 = 3, we can make a reliable estimate by adding the error (3) divided by twice our guess (4) giving us 4 + 3/8 = 4.375.

With this methodology, you can work out the square root estimate of any figure regardless of its magnitude.

Interesting Rules in Calculations

To give you the practical side of these calculations, consider the following rules:

The Rule of 70

This rule is used in estimating the number of years it takes for your investment to double. The simplest way to find the number of years is to divide 70 by the rate of interest. If the rate is 5% per annum, it will take 70 divide by 5 years to double your investment which is 14 years.

The Rule of 110

This rule is used to determine how long your investment will take to triple. As with the rule of 70, all you need here is to divide 110 by the rate of interest. Assuming the rate of interest was 10 percent per annum, it will take 11 years to triple your money.

Both of these rules respect the time value of money and will give you a solution based on the compounding effects of interest rates.

6: Speeding Up Pencil and Paper Calculations

Since the coming of calculators, the art of pencil and paper arithmetic has been pushed to the sidelines. As a result, it is slowly becoming a lost art as many of the generations prefer faster calculations using computer programs and apps. In this chapter, we will focus on ways you can speed up the pencil and paper calculations by combining them with mental computations.

Column of Numbers

In everyday life, we meet up with column of numbers either at home when doing budgeting, or at work when adding revenue and expenditures that establish the profitability of a business. Assume you want to add the following numbers:

   4328

   884

   620

   1477

617

+725

8651

The approach that is commonly used is to get a piece of pen and paper and then add from right to left. This is what we were taught in school. However, in order to do these calculations in your head faster than you would with a calculator, you should do the following:

When adding the numbers on the last column, make summations in your head as you go along. For instance, instead of 8 + 4 + 0 + 7 + 7+5, you can approach it as 8, 12, 19, 26, 31. This becomes easier because you know what to put at the end that is 1 and carry 3 to the next column. You should reverse the order and add from bottom to top because. This way, you are less likely to make a mistake.

Modular Arithmetic

Also known as the mod sums method, this technique is used to countercheck whether an answer is correct. All you need to do is to add the digit of each number until you are left with only a single digit. For instance, in computing the mod sum of 4328, take 4 + 3 +2 + 8 = 17. Thereafter, add the digits in 17 to get 1 + 7 = 8. Therefore the mod sum of 4328 is 8.

In verifying the answer 8651 in the additions done under the column of numbers, you can do the following.

Find the mod sum of 8651 which is 2.

Then find the mod sum of each of the numbers in the column. Their mod sums should correspond to this:

4328…………8

884……………2

620……………8

1477…………1

617……………5

725……………5

The summation of the individual mod sums is 29 and the mod sum of 29 is 2 which is the same as the mod sum of the solution, 8651.

Paper Subtractions

When subtracting numbers in a column, it is obvious that you cannot subtract them the same way you would add them. Subtraction is normally done number by number, meaning that at any one particular time you can only handle two numbers.

In subtractions, one way to check the accuracy of your answer is by adding the answer to the second number to see whether you will arrive at the top number.

You may also use mod sums to check the accuracy of your answer. Unlike addition, here you subtract the mod sums you arrive at and thereafter, compare the number to the mod sum of your answer. Take for instance:

53771 -------------------- 5

-34819 -------------------- -7

18952 -------------------- - 2

Where the difference in mod sums of the individual numbers is negative or zero, add 9 to it, and then compare it with the mod sum of your answer. If you add 9 to -2, you will get 7 which is the same as the mod sum of 18952.

Multiplication on Pen and Paper

The crisscross method is normally used to do multiplication without the need to write any partial results. Here is how to go about it; 47 x 34 = 1598.

The first step is to multiply the 4 in 34 and the 7 in 47 which will give you 28. Then right down the 8 and carry the 2 mentally to the next computation.

In the second step, do a crisscross where the 4 in 47 is matched with the 4 in 34 and the 3 in 34 is matched with the 7 in 47. The calculation is then done as 2 + (4x4) + (3x7) = 39. Carry the 3 in 39 and use it in the next step.

The third and last step is to multiply the first digits of the two numbers that is 3 and 4. This is how it is done; 3 + (3x4) = 15.

From the time you started, you wrote 8 at the far right followed by 9 and then 15 at the far left. This gives you the figure 1598 which is the answer.

To cross check the accuracy of your answer with the mod sum, you need to multiply the mod sums of the two respective numbers and compare it with the mod sum of the product.

In the example above, the mod sum of 47 is 2 and that of 34 is 7. Their product is 14 whose mod sum is 5. On the other hand, the mod sum of the answer (1598) is 5. This means the answer is correct.

You can take on bigger problems and still get accurate answers using this methodology.

7: Perfecting the Art of Memorizing Numbers

People who are good at numbers are often mistaken as geniuses who have an extraordinary memory. The most shocking thing, however, is that these people have a same memory like yours and mine but they have learned a system that helps them calculate their numbers right.

Again, the same memory system can be learned by everyone. According to research, any person of average intelligence can be trained to improve his ability to memorize numbers.

Memorizing numbers help you to remember dates with accuracy and even recall people's phone numbers without looking at the phone book.

Applying Mnemonics

Mnemonic is a tool that is used to enhance memory retrieval and coding. How it works is very interesting. It converts incomprehensible data into something which is more sequential and meaningful. For instance, if you want to memorize the following sentence:

My beach ball cans, fireworks blast, get my keys home.

If you can read this sentence several times and then try it on your own without looking at it, with time you will be able to visualize the beach ball, the blasting fireworks and the keys to your home.

This is what mnemonics is all about; you can use it to memorize about the first 20 digits in pi (π). These digits have no correlation or pattern at all and are usually approximated as 3.142.

The Phonetic Code

The phonetic code consists of numbers between 0 and 9 where each number is assigned a consonant sound. This is illustrated below:

1 is d or t sound

2 is n sound

3 is m sound

4 is r sound

5 is l sound

6 is ch, sh or j sound

7 is k or g sound

8 is f or v sound

9 is b or p sound

0 is s or z sound

Memorizing these sounds is not as difficult because in instances where a number is associated with more than one sound, the pronunciations are similar.

For instance, the k sound in cat and kite is similar to the g sound in goat. You can convert the numbers into sounds by adding vowel sounds around the consonants. For instance, go will be represented by 7 while maze will be represented by 30.

Where you combine more than two consonant sounds such as turtle, you need to use the respective numbers which are 1415.

The aim of phonetics is to jog your memory and bring coordination between the letters and numerics. You will be able to recall specific numbers. For instance, social security numbers, driver's license numbers, phone numbers and even the digits of pie.

Remember, the g sound in ginger is softer compared to the g in grass. Therefore ginger takes j and grass takes g.

Mnemonics and Mental Calculations

Mnemonics not only improve your capacity to memorize long numbers and sequences but it can also be used to enable storage of partial results amidst a complex mental calculation.

For instance, to store a number like 115200, you can start by storing 200 through the raising of your two fingers and then converting the 115 into a word such as title. Memorizing of numbers using the mnemonic technique gets better with practice and in no time you should be able to convert in both ways, number to words and words to numbers.

Without the use of mnemonics, the human memory has been found to hold a maximum of about 7 to 8 digits at a time. This is why you need this technique because it expands your memory considerably. It is possible to repeat a series of up to 16 digits without even looking at the board or the surface on which they are written.

8: Stretching Your Mind with Advanced Multiplication

Having gone through mental addition, multiplication, subtraction, division and the making of estimates, you have learned a lot. From here, you will be able to improve your mental math significantly. As it is commonly said, learning mathematics has no conclusive end and new things keep on coming every now and then.

This chapter is more of a continuation to what you have learned so far but be prepared to go to the deeper end. The advanced multiplications covered here will stretch your mind to the limits of mental computations.

If you are comfortable and your speed in using phonetic code is reasonably fast, then advanced multiplication will look much simpler to you. Remember, all mental computation skills can be mastered as long as you simplify the problems into easier and more manageable parts.

Four Digit Squares

In order to master four digit squares, you should be able to comfortably handle 4 by 1 multiplication problems. Take an example below:

4867 (4800 + 67) x 9 = (9 x 4800) + (67 x 9) = 43200 + 603 = 43, 803

Let us now square a four digit number 4267. The method here is used the same as two digit and three digit squares. Round down 4267 to 4000 by subtracting 267 and add the 267 to 4267 so as to get 4534.

The two numbers to be multiplied are therefore, 4534 x 4,000 which give you 18, 136, 000. To this number add the square of 267, which can be further be broken down into 267 + 33 and 267 – 33 to give you 300 and 234.

Multiply 300 by 234 to give you 70,200. To this, add the square of 33 which is 1089. The answer, therefore, is 18, 136, 000 + 70,200 + 1089 = 18, and 207,289.

To memorize, hold onto 18 million which is easy to remember then use mnemonics to memorize 136 which you can give it the code name damage (d=1 m= 3 j=6). This will help you to hold on to the figure of 18, 136,000 as you go on with your calculations.

Three by Two Multiplications

You can solve these problems using the factoring method where you break down one of the numbers. For instance in 637 x 56, you can break down the 56 into 7 x 8 and then proceed with the calculation.

In the event you come across a number with a factor of 11, take advantage of it and make it the first computation. For instance, in 462 x 52, instead of breaking up 52 into 4 x 13, you can take the advantage of breaking up 462 into 11 x 7 x 6 so that you can multiply easily with 11.

If the factoring method cannot break down the two and three digit numbers, you can use the addition method which breaks up the figures as follows:

In 761 x 37, you can break up the 761 into 760 + 1 then multiply it by 37.

Lastly, you can use the subtraction method to solve the three by two multiplication problem because sometimes it is more convenient to subtract than to add. For instance, in 758 x 43, it is convenient to make 758 as 760 – 2 instead of 750 + 8.

Five Digit Squares

Three by two multiplications require a fair amount of practice but the moment you master it, you can easily do five digit squares because they easily break up into three by two problems with a three digit and two digits square. Take the following example:

467922; this can be broken up into (46000 + 792) x (46000 + 792).

Then using distributive law, a2 +2ab + b2 where a is 46,000 and b is 792; you can further arrange it as 46,000 x 46, 000 + 2 (46,000 x 792) + (792 x 792).

Then simplify it further to 462 x 1 million + 2000 (46 x 792) + 7922

Starting with the middle section, [792 (800 - 8) x 46] x 2000 to give you 36,432 x 2000 = 72,864,000. Then the square of 46 x 1 million is 2, 116, 000, 000.

For the 7922, you solve it just like any other three digit square to give you 627, 264.

The total summation is therefore 72, 864, 000 + 2, 116, 000,000 + 627,264 = 2,189,491,624.

You can still use phonetic codes to remember the figures at every stage.

Other problems which follow the same path are three by three and five by five multiplications. The rule is the same and you can use factoring, addition or subtraction.

9: Understanding the Magical Secrets of Arithmetic

Playing around with numbers is interesting and can bring great joy in life. In fact, the more you become engrossed with arithmetic, the more entertaining it becomes.

However, for you to fully enjoy and understand the magical secrets behind numerics, you need have knowledge in algebra. Algebraic expressions have been used in many real world scenarios including computer programming, modeling and many others.

Algebra becomes interesting from the moment you appreciate the mathematical magic tricks that it comes with. This chapter will show you all that and much more.

Psychic Math

Psychic math is particularly intriguing because it is performed with a participative audience. For instance, you can tell someone to think of any number either a one digit or two digit number. Then, double the number and add 12 then divide the sum by 2 and subtract the original number. You will find that the resulting number is 6 regardless of how many digits it has.

The Magic 1089

This magic has been practiced for centuries. Let one member of your audience take out a pencil and a piece of paper then secretly jot down any three digit number where the digits are decreasing such as 973, 851, 642 etc. Then, let them reverse the number and subtract it from the original number. When you add the answer you get to the reverse of itself, you end up with 1089.

The Trick of the Missing Digit

You can pick the same number as above 1089. Give it to a volunteer and ask her to multiply it by a three digit number she prefers and she should not tell you what the number is.

After she multiplies, ask her the number of digits in the answer. It will be 6. If you ask her to call out 5 of the 6 digits in no particular order, you can try to determine the missing digit. Assuming she calls 2, 4, 7, 8, 8; you can correctly tell her that the missing digit is 7.

The secret behind this trick is that a number is a multiple of 9 if the digits sum adds up to multiples of 9. Since the sum of digits in 1089 is 18, it is also a multiple of 9. Taking 1089 and then multiplying it by any digit, the resultant answer will also be a multiple of 9.

When the digits are called up, they should add to a multiple of 9. In the above example, the numbers were adding up to 29 which is not a multiple of 9. To make it a multiple of 9, you have to add 7 to make it 36 hence the missing digit is 7.

Leapfrog Addition

This is a trick that combines prediction and mental calculation. Give someone a card that has 10 lines which are numbered from 1 to 10 and then ask them to think of any two positive numbers that lie between 1 and 20.

Let them enter the chosen numbers on line 1 and 2. Then let them write the sum of the lines 1 and 2 on line 3. Next step is to let them write the sum of lines 2 and 3 on line 4 and continue in that fashion until they get to line 10.

When the person shows you the card, you can tell him the summation of all the numbers written on the card. The answer is usually the multiplication of the number on line 7 by 11.

Quick Cube Roots

Ask anyone to pick a two digit number and then secretly cube it. When he tells you the answer, you will be able to reveal the original secret number. The only thing is to learn the cube roots of 1 to 10 then everything else becomes pretty simple.

For instance, in finding the cube root of 314, 432, you can begin with looking at the magnitude of the number in the thousands. In this case, it is 314. Since 314 lies between the cube of 6 which is 216 and 7 which is 743, the cube root must be lying in the 60s. The first digit is therefore 6.

In determining the last digit of the cube root, you should note that only 8 has a cube that ends in 2 so the last digit must be 8 and the number is 65.

Simplified Square Roots

In addition to the cube roots above, square roots can also be calculated easily as long as you are given the perfect square. For example, if you are given 7569, you can immediately tell that the square root is 87. This is how you go about it:

Analyze the magnitude of the hundreds number which in this case is 75. From simple arithmetic, you know that 75 lies between the square of 8 which is 64 and 9 which is 81.

This automatically tells you that the square root is in the 80s and the first digit is therefore 8. Since the ending digit of the square is 9, you look for the numbers whose square is 9.

Unfortunately, the numbers are two; 3 and 7. So the square root can either be 83 or 87. To find out the correct square root, you can compare the original number with the square of 85.

The reason you have picked 85 is because it is easy to compute. 85 x 85 = (90 x 80) + (5 x 5) = 7225. Since the original square is 7569, the answer must therefore be above 85 and in this case it is 87.

Other interesting tricks include the magic square and picking a day for any date.

Conclusion

Mathematics is an elegant, wonderful and exceedingly useful language. It has its own nouns, verbs, modifiers, patois and dialect. While some people use this language brilliantly, others are not as enthusiastic. This is because many fear the numbers and how complex they can get.

The contents in this book may not guarantee you 100 percent that you will be a professor of algebra the moment you turn the last leaf. However, it is bound to bring a new, exciting and entertaining perspective of the potential that lies in numbers.

Different people have different levels of confidence when it comes to arithmetic computations and for some reason; others feel comfortable with what little they know. In case they are confronted with complex problems, they do not hesitate to reach out to their pocket calculators.

Just like photography which may blind us to the beauty of a Vermeer painting or a keyboard that may drown the magnificence of a sonata, over-reliance on automated calculations can deny us the beauty that is hidden in arithmetic. This book has been written in a very illustrative manner which makes math fun to learn and internalize.

Every formula and trick included can be learned as long as you have the interest. You do not need a super mind for you to improve your math skills but rather your curiosity.

Even if you retain and use only a few of the concepts discussed in this book, it will still have fulfilled its purpose in you. It is worth the investment of your time.

Made in the USA
San Bernardino, CA
14 July 2015